Flavonoids of Pepper as Cytokines Inhibitors

Ahmed Ezzaldean Abdellah Allam

ELIVA PRESS

ELIVA PRESS

Ahmed Ezzaldean Abdellah Allam

Various naturally occurring phytochemicals exhibit anti-inflammatory activity and are considered to be potential drug candidates against inflammation-related pathological processes. Here, new flavonoid glycoside, kaempferol 3-O-α-¬[(6-P-coumaroyl galactopyranosyl-O-β-(1→4)-O-α-rhamnopyranosyl-(1→4)]-O-α-rhamnopyranoside 1, in addition to five known flavonoid glycosides (2-6) kaempferol 3-O-[α-¬rhamnopyranosyl-(1→4)-O-α-rhamnopyranosyl-(1→6)-O]-β-galactopyranoside (kaempferol 3-O-β-isorhamninoside) 2, quercetin 3-O-[(2,3,4-triacetyl-α-rhamnopyranosyl)-(1→6)-β-galactopyranoside 3, quercetin 3-O-[(2,4-diacetyl-α-rhamnopyranosyl)-(1→6)]-3,4-diacetyl-β-galactopyranoside 4, quercetin 3-O-[(2,4-diacetyl-α-rhamnopyranosyl)-(1→6)]-2,4-diacetyl-β-galactopyranoside 5, quercetin 3-O-[(2,3,4-triacetyl-α-rhamnopyranosyl)-(1→6)-3-acetyl-β-galactopyranoside 6 were isolated from bell pepper (Capsicum annum L.) fruits and tested for both anti-inflammatory activity through cytokine production (TNF-α and IL-1β) and antioxidant activity through scavenging effect on 1,1-diphenyl-2-picrylhydrazyl (DPPH) assay. Compounds 1-3 significantly suppressed production of TNF-α / IL-1β in cultured THP-1 cells previously co-stimulated by LPS in a dose-dependent manner (10.2/49.1, 28.1/55.7, and 35.2/57.5 μM respectively) whereas compounds 4-6 have relatively weaker inhibitory activity. (45.3/73.5, 48.2/65.6, and 42.2/67.4 μM respectively). All compounds 1-6 showed no cytotoxic activity against the growth of THP-1where the percentage of cell viability was (127.4, 108.5, 105.4, 103.9, 103.4, and 104.2 μM respectively). All isolated compounds exhibited higher radical scavenging activity than ascorbic acid in (DPPH) assay. These results indicated that bell pepper fruits could be an effective candidate for ameliorating inflammatory-associated complications.

Published by Eliva Press SRL
Address: MD-2060, bd.Cuza-Voda, 1/4, of. 21 Chişinău, Republica
Moldova
Email: info@elivapress.com
Website: www.elivapress.com

ISBN: 978-1-63648-182-1

Flavonoids of Pepper as Cytokines Inhibitors

Ahmed E. Allam*

*Department of Pharmacognosy, Faculty of Pharmacy, Al-Azhar University

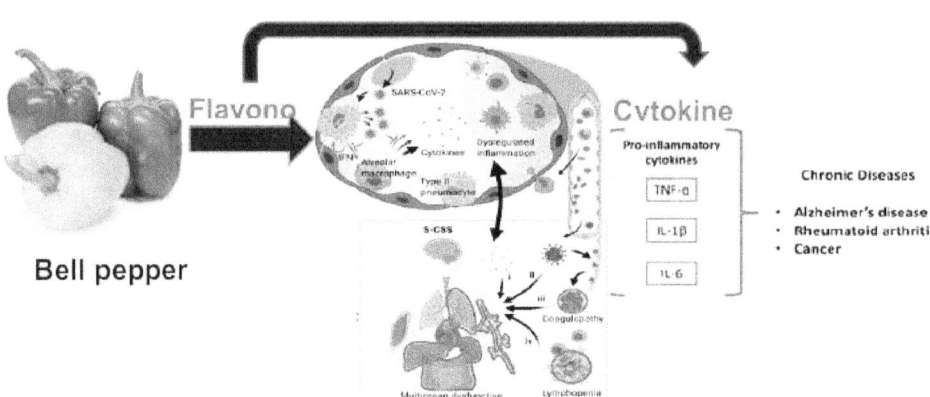

Bell pepper Flavono Cvtokine

Pro-inflammatory cytokines

TNF-α

IL-1β

IL-6

Chronic Diseases

- Alzheimer's disease
- Rheumatoid arthritis
- Cancer

Table of contents

ABSTRACT

Various naturally occurring phytochemicals exhibit anti-inflammatory activity and are considered to be potential drug candidates against inflammation-related pathological processes. Here, new flavonoid glycoside, kaempferol 3-*O*-α-[(6-P-coumaroyl galactopyranosyl-*O*-β-(1→4)-*O*-α-rhamnopyranosyl-(1→4)]-O-α-rhamnopyranoside **1**, in addition to five known flavonoid glycosides (**2-6**) kaempferol 3-*O*-[α-rhamnopyranosyl-(1→4)-*O*-α-rhamnopyranosyl-(1→6)-*O*]-β-galactopyranoside (kaempferol 3-*O*-β-isorhamninoside) **2**, quercetin 3-*O*-[(2,3,4-triacetyl-α-rhamnopyranosyl)-(1→6)-β-galactopyranoside **3**, quercetin 3-*O*-[(2,4-diacetyl-α-rhamnopyranosyl)-(1→6)]-3,4-diacetyl-β-galactopyranoside **4**, quercetin 3-*O*-[(2,4-diacetyl-α-rhamnopyranosyl)-(1→6)]-2,4-diacetyl-β-galactopyranoside **5**, quercetin 3-*O*-[(2,3,4 -triacetyl- α-rhamnopyranosyl)- (1→6)-3- acetyl- β-galactopyranoside **6** were isolated from bell pepper (*Capsicum annum* L.) fruits and tested for both anti-inflammatory activity through cytokine production (TNF-α and IL-1β) and antioxidant activity through scavenging effect on 1,1-diphenyl-2-picrylhydrazyl (DPPH) assay. Compounds **1-3** significantly suppressed production of TNF-α / IL-1β in cultured THP-1 cells previously co-stimulated by LPS in a dose-dependent manner (10.2/49.1, 28.1/55.7, and 35.2/57.5 µM respectively) whereas compounds **4-6** have relatively weaker inhibitory activity. (45.3/73.5, 48.2/65.6, and 42.2/67.4 µM respectively). All compounds **1-6** showed no cytotoxic activity against the growth of THP-1where the percentage of cell viability was (127.4, 108.5, 105.4, 103.9, 103.4, and 104.2 µM respectively). All isolated compounds exhibited higher radical scavenging activity than ascorbic acid in (DPPH) assay. These results indicated that bell pepper fruits could be an effective candidate for ameliorating inflammatory-associated complications.

Keywords: Anti- inflammatory; Antioxidant activity; Bell pepper; Cytokine production

1. Introduction

Because the discovery of new drugs requires a long time and expense, searching for new compounds from natural sources known for their high safety and applicability will be a good avenue to deal with inflammatory-associated complications. From which, bell pepper (*Capsicum annum* L.), a member of the family Solanaceae is the most commonly consumed fruity vegetable well known for its nutritional and antioxidant significance [1]. Bell pepper exhibits medicinal as well as food value all over the world. Both warm and dry climates are suitable for its cultivation [2]. It is a rich source of neutral phenolic compounds, especially luteolin, quercetin, capsaicinoids, nordihydrocapsaicin, homodihydrocapsaicin, homocapsaicin, norcapsaicin, and nornorcapsaicin. Capsaicin possesses noteworthy analgesic and anti-inflammatory properties. It also possesses healing effects for treatment of arthritis, diabetic neuropathy, gastric lesions, and cardiac excitability that is why it is incorporated in creams and gels [3, 4]. Carotenoids exhibit antioxidant properties and prevent tissue damage by acting as singlet molecular oxygen; reactive oxygen species (ROS); peroxyl radicals and reactive nitrogen species (RNS) scavengers [5]. Red peppers have been used for several thousand years as food additives and for a broadvariety of medical applications in Indian, Native American, African and Chinese medicaltraditions (Govindarajan & Sathyanarayana 1991; Szallasi & Blumberg 1999). Red peppers have been claimed to enhance immune response, act in an anti-inflammatorymanner, lower blood pressure, reduce excessive blood clotting and reduce blood sugar levels,but no formal examination of these claims has been published. Capsaicin is the pungent component of red peppers and

6

because of its analgesic and anti-inflammatory activity has been used in clinical practice. Thus, topical application of capsaicinhas a therapeutic value in a variety of neuropathic pain conditions, including rheumatoidarthritis, osteoarthritis, diabetic neuropathy, postmastectomy pain syndrome, psoriasis, burn-ing mouse syndrome and herpes zoster. Immunosuppressive drugs are used for the treatment of undesirable or abnormal activation of T lymphocytes and the immune system associated with organ transplantation and autoimmune diseases. T lymphocytes play a pivotal role in the pathogenesis of cell-mediated autoimmune diseases and chronic inflammatory disorders (1–3). Activation of T lymphocytes requires stimulation of T-cell receptors and costimulatory signals. Ca2+ influx is crucial for T-cell activation upon antigen stimulation (4). The best known costimulatory signals are those of NF-κB (nuclear factor kappa-light-chain-enhancer of activated B cells). NF-κB is activated by Ca2+/calmodulin dependent protein kinase (CaMK II) (5,6). NF-κB translocates to the nucleus and turns on transcription of specific genes, generally related to inflammatory or immune responses, cell survival responses, or cell proliferation.

Recently, fruits and vegetables have been recognized as natural sources of various bioactive compounds (7–9). Natural antioxidants often exist together in different combinations in nature, and consequently researchers have been investigating the additive and synergistic effects of different antioxidants

Inflammation can be a blessing or a blight. It is a critical part of the body's immune response that in normal circumstances reduces injury and promotes healing. When it goes awry, however, the inflammatory response can lead to serious physical and mental problems. Inflammation plays a key role in many neurodegenerative diseases and also is implicated in the cognitive and behavioral impairments seen in aging.

Some studies looked at luteolin (LOO-tee-OH-lin), a plant flavonoid known to impede the inflammatory response in several types of cells outside the central nervous system. "One of the questions we were interested in is whether something like luteolin, or other bioactive food components, can be used to mitigate age-associated inflammation and therefore improve cognitive function and avoid some of the cognitive deficits that occur in aging," Johnson said.

The researchers first studied the effect of luteolin on microglia. These brain cells are a key component of the immune defense. When infection occurs anywhere in the body, microglia respond by producing inflammatory cytokines, chemical messengers that act in the brain to orchestrate a whole-body response that helps fight the invading microorganism.

This response is associated with many of the most obvious symptoms of illness: sleepiness, loss of appetite, fever and lethargy, and sometimes a temporary diminishment of learning and memory. Neuroinflammation can also lead some neurons to self-destruct, with potentially disastrous consequences if it goes too far.

Graduate research assistant Saebyeol Jang studied the inflammatory response in microglial cells. She spurred inflammation by exposing the cells to lipopolysaccharide (LPS), a component of the cell wall of many common bacteria.

Those cells that were also exposed to luteolin showed a significantly diminished inflammatory response. Jang showed that luteolin was shutting down production of a key cytokine in the inflammatory pathway, interleukin-6 (IL-6). The effects of luteolin exposure were dramatic, resulting in as much as a 90 percent drop in IL-6 production in the LPS-treated cells.

Using electromobility shift assays, which measure the binding of transcription factors to DNA promoters, Jang eventually determined that luteolin inhibited IL-6 production by preventing activator protein-1 (AP-1) from binding the IL-6 promoter.

AP-1 is in turn activated by JNK, an upstream protein kinase. Jang found that luteolin inhibited JNK phosphorylation in microglial cell culture. The failure of the JNK to activate the AP-1 transcription factor prevented it from binding to the promoter region on the IL-6 gene and transcription came to a halt.

To see if luteolin might have a similar effect in vivo, the researchers gave mice luteolin-laced drinking water for 21 days before injecting the mice with LPS.

Those mice that were fed luteolin had significantly lower levels of IL-6 in their blood plasma four hours after injection with the LPS. Luteolin also decreased LPS-induced transcription of IL-6 in the hippocampus, a brain region that is critical to spatial learning and memory.

The findings indicate a possible role for luteolin or other bioactive compounds in treating neuroinflammation,

So, It might be possible to use flavonoids to inhibit JNK and mitigate inflammatory reactions in the brain. Inflammatory cytokines such as interleukin-6 are very well known to inhibit certain types of learning and memory that are under the control of the hippocampus, and the hippocampus is also very vulnerable to the insults of aging.

One such vegetable where a variety of antioxidants can be found is the pepper. The pepper belongs to the genus Capsicum, which contains >200 varieties, with Capsicum annuum, Capsicum baccatum, Capsicum chinense, Capsicum frutescens, and Capsicum pubescens being the main five species. Peppers are consumed worldwide and their importance has gradually increased to place them among the most consumed spice crops in the world. They also have a significant role in traditional medicine. It has been reported that the red pepper fruit of Capsicum baccatum shows anti-inflammatory activity via nitric oxide scavenging activity. These findings indicate that the bell pepper has a potential immunosuppressive function through its effects on cells. However, the anti-inflammatory and underlying immunosuppressive mechanisms of the bell pepper (Capsicum annuum L. var. grossum), one of the species of the genus Capsicum, are still largely unknown and require further investigation. In this study, we investigated the in vitro anti-inflammatory effect of water extract from bell pepper leaves (WEBP) on mouse spleen cells, and explored the potential mechanism underlying this effect. We found that WEBP significantly inhibited Con-A-stimulated spleen cell proliferation, cytokine production, and expression of inflammatory proteins. The data also showed that WEBP exhibited an immunosuppressive effect, via inhibition of T-cell activation through the NF-κB pathway.

A cytokine is a protein secreted by cells that play a role in immune response management, so by suppressing cytokines, pepper can help turn down the volume on inflammatory reactions. The utilization of bioactive compounds plays key health-promoting functions such as protecting against oxidative cell damage, cancer insurgence, diabetes prevalence, cardiovascular disorders, Alzheimer's, and Parkinson's disease [6,7] that evoked us to carry out this study.

2. Materials and methods

2.1. General

The optical activity was determined with Horiba SEPA-3000 high-sensitivity polarimeter (Horiba). UV analysis was determined on Shimadzu UV-1600 UV-visible spectrometer, IR analysis was operated on Shimadzu FTIR-8400. NMR analysis was carried out on JEOL GSX-600 spectrometer in CD_3OD. Fast atom bombardment (FABMS) and high-resolution fast atom bombardment (HRFABMS) were carried out on JEOL JMS SX-102 mass spectrometry. Reversed-phase high-performance liquid chromatography (HPLC) was undertaken on an ODS column (particle size: 5 μm, TOSO, 18 × 250 mm) RP-23 (5 μm; Waters). Diaion HP-20 (Mitsubishi) (Tokyo Japan), silica gel (63-210 μm; Kanto Kagaku), and ODS (63-212 μm; Wako Pure Chemical) (Tokyo Japan) were used for open column chromatography. Thin-layer chromatography (TLC) was carried out on silica gel (SiO2, 60-100 mesh; Wako Pure Chemical) 60 F254 and RP-18 F254S (Merck).

2.2.1. Extraction and Isolation

Air-dried bell pepper fruits (2 kg) were extracted thrice with MeOH (5 L each) to yield methanol extract (310 g) which was partitioned between distilled water, chloroform, ethyl acetate, and n-butanol (1 L each) to yield chloroform fraction (90 g), ethyl acetate fraction (60 g), the n-butanol fraction (50 g) and the rest aqueous fraction (100 g). All fractions were screened for the antioxidant and cytokine production in cultured THP-1 cells activities where the ethyl acetate was the most active fraction and hence, and hence it was fractionated by ODS column using six mobile phase systems of CH_3CN-H_2O (10, 25, 40, 50, 70 and 90% v/v; elution volume: 500 ml of each) to give six corresponding subfractions. Subfraction eluted with 40% CH_3CN (3.8 g) was further isolated by silica gel column chromatography with gradient elution by $CHCl_3$:MeOH (ratios of 9:1, 6:1, 4:1, 3:1 and 1:1, v/v, elution volume: 200 ml each) to give five corresponding subfractions. The subfraction eluted by 6:1 $CHCl_3$: MeOH was further chromatographed by preparative HPLC, ODS column equipped with a UV detector (210 nm) with mobile phase 20% CH_3CN in H_2O which afforded compounds **3-6**, (15, 18, 22, and 9 mg respectively). These preparative HPLC conditions were also used after gradually increasing the mobile phase to 50% CH$_3$CN in H$_2$O to isolate the same fraction to afford compounds **1** and **2**, (13 and 24 mg respectively).

2.2.2. Acid Hydrolysis

Acid hydrolysis of the flavonoid glycosides was carried out by refluxing 5 mg of the compound in 5 ml of 6% HCl in MeOH for 3 h. The reaction mixture was partitioned against EtOAc (3 × 10 ml). The aglycones were obtained from the EtOAc layer and identified as kaempferol and quercetin by co-chromatography on silica gel with reference samples (Sigma) (Tokyo, Japan). Identification of galactose and rhamnose present in the sugar fraction was carried out by comparison with authentic samples, galactose (Rf 0.41), glucose (Rf 0.46), and rhamnose (Rf 0.66) (Sigma) (Tokyo, Japan) in TLC on silica gel (CHCl$_3$-MeOH-H$_2$O 8:5:1) using 5% H$_2$SO$_4$ in MeOH as spraying reagent followed by heating the plates at 120° C for 15-20 min.

2.2.3. Measurements of the optical rotation of D-galactopyranose and L-rhamnopyranose tetrabenzoate derivatives

Benzoyl chloride (0.5 ml) was added to each ice-cooled solution of either D–galactopyranose (6.0 mg) or L-rhamnopyranose (4.0 mg) in dry pyridine (1.0 ml) and each mixture was stirred at room temperature for 15 h. MeOH (1.0 ml) was added dropwise to the reaction mixture, stirred for 30 min, and then diluted with EtOAc and aqueous Na$_2$CO$_3$, and the layers were separated. Each organic layer was washed with brine and the combined aqueous layers for each was extracted with EtOAc. Each combined organic extract was dried over MgSO4 and concentrated. The corresponding residual dark brown oil fractions were individually purified by silica gel cc (eluted by hexane/EtOAc 5:1) to give either D–galactopyranose

12

tetrabenzoate $[\alpha]_D^{31}$ + 53.5 (c= 1.2, CHCl$_3$) or L-rhamnopyranose tetrabenzoate $[\alpha]_D^{29.6}$ + 75.0 (c =1.6, CHCl$_3$) as a colorless oil, respectively [8,9].

2.3.1 Evaluation of cytokine production in cultured THP-1 cells and cytotoxic assay

To determine the effect of both fractions and the isolated compounds (1-6) on the production of inflammatory cytokines in monocytes THP-1 cells (Dainippon Pharmaceutical Company), a method modified by (Bornstein et al; 2004 and Nehmé et al; 2008) was used [10, 11].

2.3.2. MTT in Vitro Assay

To determine the cytotoxic activity of the tested samples, THP-1 cells (180 μL) were seeded in 96-well plates at 1.0×105 cells per well with tested samples (purity > 93%) (20 μL in DMSO/ PBS) at various concentrations. After 48-h cultivation, supernatants were removed, non adherent cells (THP-1) incubated with 3-(4, 5-dimethylthiazol-2-yl)-2,5-diphenyltetrazolium bromide (MTT; 10 μL, 5 mg/mL in PBS) for 4 h, and then solubilized with 10% (w/v) sodium dodecyl sulfate (SDS; in 60% [v/v] dimethyl formamide) solution (100 μL) for 18 h. The absorbance was measured at 570 nm using a microplate reader, and the cytotoxicity was calculated by comparing absorbance with that of the non-treated control culture. The cell growth curve was graphed using statistical analysis software (Kaleida Graph version 4.00; Synergy Software), and IC50 values calculated using simple linear regression. The cytotoxic activity of all of isolates was determined by MTT colorimetric assay (Segun, et al., 2019; Alley et al., 1988) [12,13].

2.3.3. DPPH Radical Scavenging Activity

DPPH assay was performed by a method previously reported by (Kumar et al; 2011) [14]. 100 µl of the tested samples at different concentrations in MeOH and 1.0× 10- 4 M DPPH in MeOH (300 µl) were added to the 96-well microtiter plate. The plate was shaken for 1 min on a plate shaker and incubated for 30 min at room temperature in the dark. After incubation, the absorbance was recorded at 517 nm. The tested samples at different concentrations without DPPH solution were used as a blank control to eliminate the influence of sample color. Ascorbic acid was used as a positive control [15] and DPPH solution in MeOH served as a negative control.

3. Results and discussion

3.1. Structure Elucidation of compound 1

Compound **1** (Table 1, Fig.1) (13 mg) was obtained as yellow amorphous powder, soluble in methanol, with $[\alpha]_D^{30.1}$ -54.5° (c = 0.333, MeOH). The structure of compound 1 was elucidated by UV, IR, one- and two dimensional NMR spectroscopy including ^{1}H, ^{13}C NMR, DEPT-135 H-H COSY, HMQC, and HMBC experiments, as well as HRFAB mass spectrometry. UV spectrum (in MeOH) exhibited absorption maxima at 256 nm (band-II) and 352 nm (band-I) indicating a flavonol type. IR spectrum of **1** (in CHCl$_3$) indicated the presence of hydroxyl (3445 cm-1), carbonyl (1782 cm-1) and phenyl (2980, 1640, 1533 cm-1). ^{1}H and ^{13}C NMR spectra of **1** indicated the presence of kaempferol moiety, three sugar moieties (one hexose unite and two pentose unites) in addition to the presence of *P*-coumaroyl moiety attached to a terminal hexose. ^{1}H NMR spectrum showed pair of doublets at δ$_H$ 6.09 (H-6) and δ$_H$ 6.29 (H-8) and two-spin system with the typical coupling pattern of 1,4-disubstituted benzene ring [a pair of doublets each is equivalent to two protons at δ$_H$ 7.97 (H-2', H-6') and δ$_H$ 6.82 (H-3', H-5')] which are two features characteristic of a flavonol with hydroxyl functionality at positions 5, 7 and 4'. ^{1}H-

14

NMR spectrum showed also signals characteristic to *p*-coumaroyl moiety, where there is a typical coupling pattern of 1,4-disubstituted benzene ring [a pair of doublets each is equivalent to two protons at δ_H 7.65 (H-2'''',H-6'''') and at δ_H 6.66 (H-3'''',H-5'''')] in addition to a trance olefinic protons at δ_H 6.86 and 5.77 (H-7,H-8) respectively. ^{13}C NMR agreed with 3-substituted kaempferol moiety. Substitution of kaempferol in C-3 was evident from the chemical shift of neighboring C-2 (δ_C 159.2 ppm) whereas in flavonols with unsubstituted hydroxyl functionality at C-2 was detected around δc 147 ppm [16]. A long-range correlation was observed in HMBC experiment between C-3 of kaempferol (δ_C 134.1 ppm) and the anomeric proton of rhamnose (δ_H 4.36 ppm) confirmed that this was the site of glycosylation and rhamnose was the first sugar. Other long-range correlations observed in HMBC experiment, between C-4 of the first rhamnose (δ_C 74 ppm) and the anomeric proton of the middle one (δ_H 5.1 ppm) and between C-4 of the middle rhamnose (δ_C 74 ppm) and the anomeric proton of the terminal galactose (δ_H 5.62 ppm) confirmed that the attachment between the middle rhamnose and first one is (1→ 4) and that between the terminal galactose and the middle rhamnose is (1→ 4). Another HMBC correlation was observed between H-6 protons of the terminal galactose (δ_H 3.41 and 3.13 ppm) and the carbonyl group of *P*-coumaroyl moiety (δ_C 167.6 ppm), confirming that the *P*-coumaroyl moiety is attached to the terminal galactose. Also, ^{13}C- downfield shift of both C-6 of galactose, C-4 of the middle rhamnose, and C-4 of the first rhamnose confirmed the mentioned site of attachment [17]. The HRMS spectrum showed a quasi-molecular ion peak at m/z 887.2629 [M + H] + calculated as 886.2532 per the molecular formula $C_{42}H_{46}O_{21}$. Hence **1** could unequivocally be identified as Kaempferol 3-*O*-α-[(6-*P*-coumaroyl galactopyranosyl-*O*-β-(1→4)-*O*-α-rhamnopyranosyl-(1→4)]-*O*-α-rhamnopyranoside which is a new compound.

In addition, five known falvonoid glycosides (**2-6**) Fig.1, Kaempferol 3-*O*-[α--rhamnopyranosyl-(1→4)-*O*-α-rhamnopyranosyl-(1→6)-*O]*-β-galactopyranoside (kaempferol 3-*O*-*β*-isorhamninoside) **2**, quercetin 3-*O*-[(2,3,4-triacetyl-*α*-rhamnopyranosyl)-(1→6)-*β*-galactopyranoside **3**, quercetin 3-*O*-[(2,4-diacetyl-*α*-rhamnopyranosyl)-(1→ 6)]-3,4-diacetyl-*β*-galactopyranoside **4**, quercetin 3-*O*-[(2,4-diacetyl-*α*-rhamnopyranosyl)-(1→6)]-2,4-diacetyl-galactopyranoside **5**, quercetin 3-*O*-[(2,3,4-triacetyl-α-rhamnopyranosyl)-(1→6)-3-acetyl-*β*-galactopyranoside **6** were isolated and their structure were identical with the reported data [18,-20].

3.2. Inhibition of Production of Inflammatory Cytokines in THP-1 Cells

Both fractions and isolated compounds (1-6) were subjected to 24-h stimulation by lipopolysaccharide (LPS) at 1μM (Table 2, 3; Fig. 2) where ethyl acetate fraction showed the strongest inhibitory activity. This fraction inhibited TNF-α production and DPPH radical scavenging activity with IC50 values of 19.9 and 21 μg/ml respectively. Its inhibitory activity was 2.5-1.5 times stronger than those of the original methanol extract.

Ethyl acetate fraction was subjected to isolation by HPLC yielding the previously mentioned six flavonoids 1-6 which in turn were subjected to the same assay as the parent fraction. Putative immunosuppressive glucocorticoid agent dexamethasone was used as a positive control for inhibition of cytokine production at 5 μM, dexamethasone strongly decreased levels of TNF-α, and IL-1β in culture supernatants of THP-1 cells co-stimulated with LPS [21]. It was noticed that compounds **1-3** strongly suppressed the production of TNF-α / IL-1β in THP-1 cells in a dose-dependent manner (10.2/49.1, 28.1/55.7, and 35.2/57.5 μM respectively) whereas compounds **4-6** have relatively weaker inhibitory activity.(45.3/73.5, 48.2/65.6, and 42.2/67.4 μM respectively), These represents results more or less more potent than the positive control Dexamethasone.

A possible mechanism for flavonoids used to alleviate some inflammation-related symptoms is to act by inhibiting cyclooxygenase activity or by blocking cytokine receptors [22, 23].

As SAR study, it was noticed that the effect of kaempferol and quercetin on cytokine-induced pro-inflammatory status of cultured human endothelial cells was always more strongly inhibited in kaempferol-treated than in quercetin-treated cells [24], Kaempferol can modulate the Th1/Th2 balance and could be useful for the treatment of cell-mediated immune diseases [25] which may explain the strong suppression activity of compounds (1-2) (kaempferol derivative) over compounds (4-6) (quercetin derivatives), for compound 3, it has a relatively weaker activity.

3.3. DPPH Radical Scavenging Activity

The isolated flavonoid compounds 1-6 exhibited free radical scavenging activity with IC50 values of 1.37 and 1.42 µM for 1and 2 (kaempferol derivatives) and 3.62, 3.29, 1.8, and 7.21 µM for 3-6 (quercetin derivatives), (Table 4; Fig.3). This may be attributed to the differences in the sites of the acetyl groups in compounds 3-6.

4. Conclusion

Based on the previous results, ball pepper fruit extracts may be safely considered potential anti-inflammatory and antioxidant candidates for inflammatory diseases.

Acknowledgments

The author acknowledges Pharmacognosy and Chemistry of Natural Products Department, School of Pharmaceutical Sciences, Kanazawa University, Kanazawa, Japan, for carrying out NMR analysis.

Declarations

Funding

This work was financially by the Egyptian Ministry of Higher Education and Scientific Research

Position	^{13}C NMR (δ, mult.)	^{1}H NMR [δ, mult, J (Hz)]	Position	^{13}C NMR (δ, mult.)	^{1}H NMR [δ, mult, J (Hz)]
2	159.2,s	-	Galactopyranosyl		
3	134.1,s	-	1''''	100.5,d	5.62,d,7.5
4	179.4,s	-	2''''	72.4,d[b]	3.82,dd, 9.9,7.5
5	163.1,s	-	3''''	73.8,d	3.87,dd, 9.9,3.4
6	99.8,d	6.09,d, 2.06	4''''	72.0,d[b]	5.20,d,3.4
7	165.7,s	-	5''''	77.7,d	3.84,m
8	94.7,d	6.29,d, 2.06	6''''	67.1,t	3.41,m,
					3.13,m
9	158.3,s	-	P-coumaroyl		
10	105.9,s	-	1'''''	127.6,s	-
1'	123.1,s	-	2'''''	116.0,d	6.66,2H,d,8..9
2'	132.2, d	7.97, 2H,d, 8.9	3'''''	134.0,d	7.65,2H,d,8.9
3'	115.9,d	6.82, 2H,d, 8.9	4'''''	160.0,s	-
4'	161.3,s	-	5'''''	134.0,d	7.65,2H,d,8.9
5'	115.9,d	6.82, 2H,d, 8.9	6'''''	116.0,d	6.66,2H,d,8..9
6'	132.2,d	7.97, 2H,d, 8.9	7	146.1,d	6.86,d,13.0
Rhamnopyranosyl			8	115.9,d	5.77,d,13.0
1''	102.7,d	4.36, brs.	C=O	167.6,s	-
2''	72.1,d[a]	3.88,dd, 3.4, 1.7			
3''	72.3,d[a]	3.39,dd, 9.6, 3.4			
4''	74.0,d	3.25,dd, 9.9, 9.6			
5''	69.9,d	4.0,dd,9.9,6.1			
6''	17.4,q	0.85,d, 6.1			
Rhamnopyranosyl					
1'''	102.6,d	5.10,brs.			
2'''	71.9,d	3.71,m			
3'''	73.7,d	3.74,m			
4'''	74.0,d	3.12,m			
5'''	69.9,d	3.29,m			
6'''	17.88,q	0.96,d, 6.1			

Table 1: ^{13}C and ^{1}H NMR assignments for compound 1 recorded in CD$_3$OD

Table 2: Effects of methanol extract and fractions isolated from a particular extract on TNF-α production in cultured THP-1

Extract/fraction	Inhibitory activity (IC50) (µg/ml)		DPPH radical scavenging activity	
	TNF-α production	Dexamethasone; positive control 5 µM	Ascorbic acid positive control 12 µM	n
Total methanol extract	52±8		32±8	3
Chloroform fraction	130±6		45±5	3
Ethyl acetate fraction	19±9		21±8	3
n-butanol fraction	75±3		65±4	3

Table 3: Activities of compounds 1-6 on the production of inflammatory cytokines and their cytotoxic activity in vitro.

Compound	Inhibitory activity (IC_{50}) (µM) Cytokines (TNF-α/IL-1β)	% of Cell viability Cell lines THP-1	n
1	10.2/49.1	127.4	2
2	28.1/55.7	108.5	2
3	35.2/57.5	105.4	2
4	45.3/73.5	103.9	2
5	48.2/65.6	103.4	2
6	42.2/67.4	104.2	2
Dexamethasone	42.1/40.5	ND	2

Values are mean values from two experiments (n = 8 in total) n: number of experiments (one experiment: n = 4)

Dexamethasone (positive control for the activity of cytokine inhibition) ND: not determined

Table 4: DPPH radical scavenging activity of compounds 1-6

Compound	IC_{50} (μM) [a]
1	1.37
2	1.42
3	3.62
4	3.29
5	1.87
6	7.21
Ascorbic acid	12

[a] IC_{50} values were determined by regression analysis and expressed as the mean of four replicates.

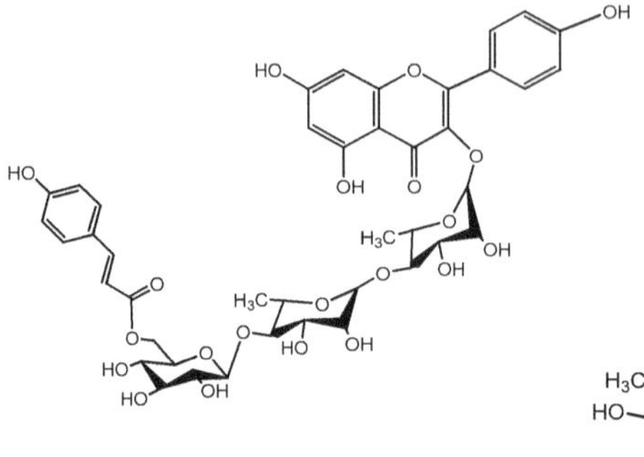

1

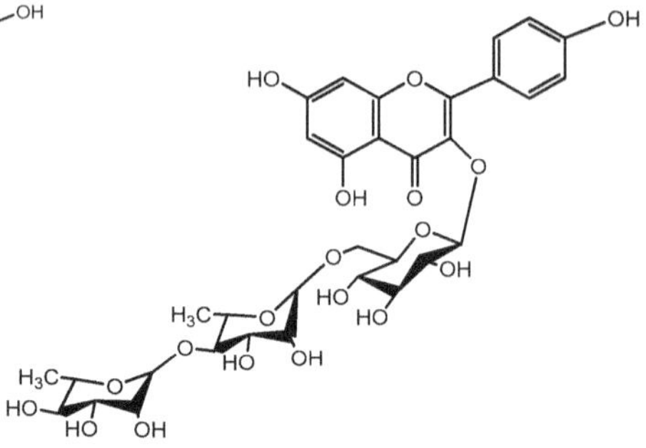

2

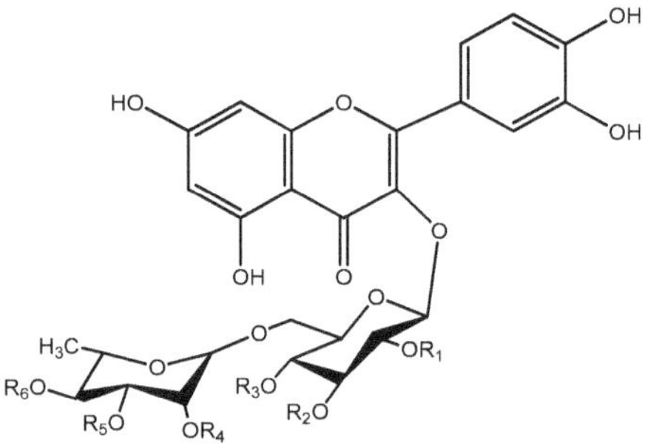

3-6

Cpd. NO.	R1	R2	R3	R4	R5	R6
3	H	H	H	CH_3COO	CH_3COO	CH_3COO
4	H	CH_3COO	CH_3COO	CH_3COO	H	CH_3COO
5	CH_3COO	H	CH_3COO	CH_3COO	H	CH_3COO
6	H	CH_3COO	H	CH_3COO	CH_3COO	CH_3COO

Fig. 1: Structure of compounds 1-6

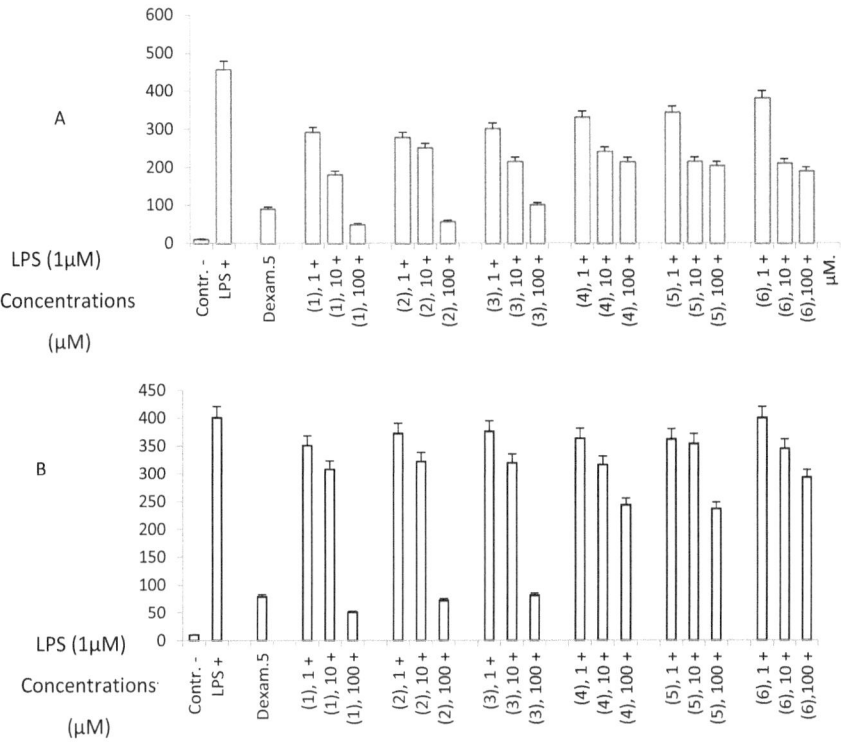

Fig. 2: Effects of flavonoid glycosides on the production of inflammatory cytokines in cultured THP-1 cells.

Cells were directly treated with various concentrations (1, 10 and100μM) of compounds **1-6** co-stimulated with 1 μM LPS. Culture supernatants were collected 24 h after LPS stimulation and cytokine levels such as TNF-α (**A**) and IL-1β (**B**) were measured using ELISA. Data are means ± SD of quadruplicate cultures. Correlation coefficients for TNF-α production co-stimulated with LPS: $r^2 = 0.89629$ (**1**), $r^2 = 0.96711$ (**2**), $r^2 = 0.92035$ (**3**), $r^2 = 0.95413$ (**4**), $r^2 = 0.99115$ (**5**), $r^2 = 0.87086$ (**6**). Correlation coefficients for IL-1β production co-stimulated with LPS: $r^2 = 0.99760$ (**1**), $r^2 = 0.99454$ (**2**), $r^2 = 0.99024$ (**3**), $r^2 = 0.90043$ (**4**), $r^2 = 0.99944$ (**5**), $r^2 = 0.95169$ (**6**).

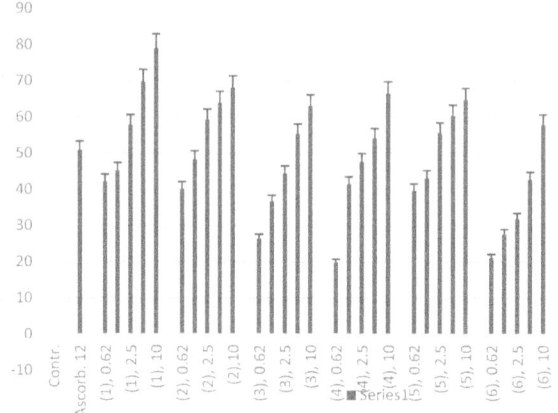

Data are means ± SD of three independent experiments (n = 12 in total); n: numbers of experiment (one experiment: n = 4). [a] Original methanol extract. [b] Original extract was loaded onto HP-20, and successive elution with water, methanol and acetone were carried out yielding three fractions

Fig. 3: DPPH radical scavenging activity of compounds 1-6

Fig. S1; ^{1}H-NMR spectrum of compound 1 (CD$_3$OD, 600Mz)

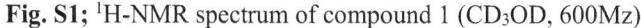

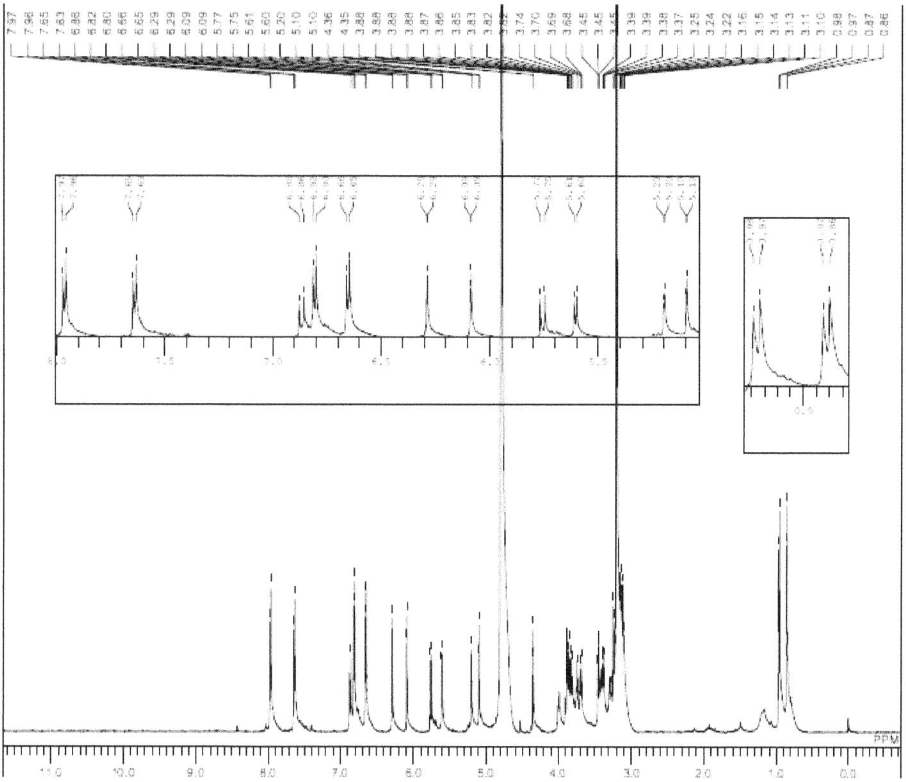

Fig. S2; ^{13}C-NMR spectrum of compound 1 (CD$_3$OD, 150Mz)

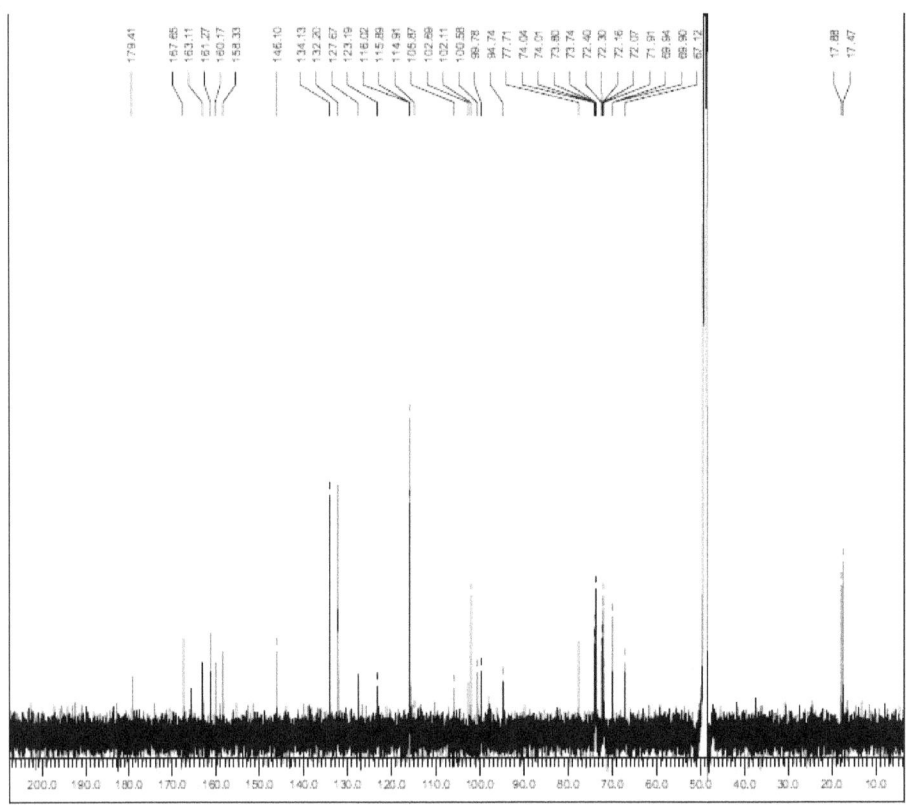

Fig. S3; DEPT 135 spectrum of compound 1 (CD$_3$OD, 150Mz)

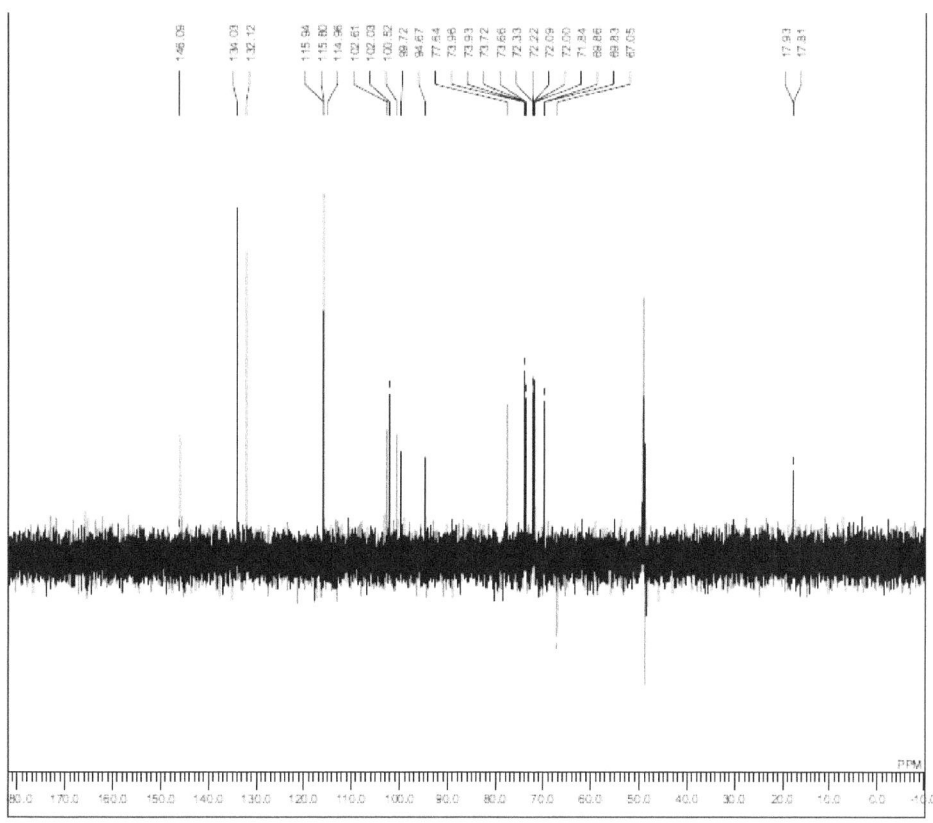

Fig. S4; H-H COSY spectrum of compound 1 (CD₃OD, 600Mz)

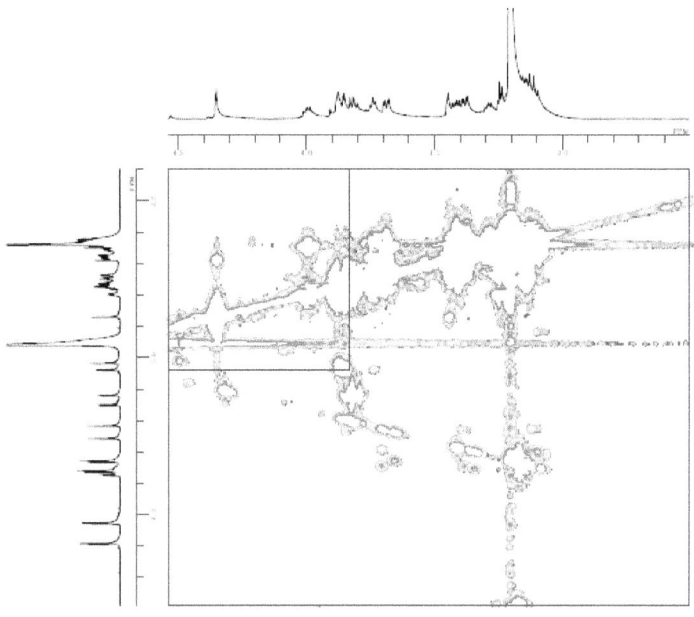

Fig. S5; HSQC correlation of compound 1 (CD₃OD, 600Mz)

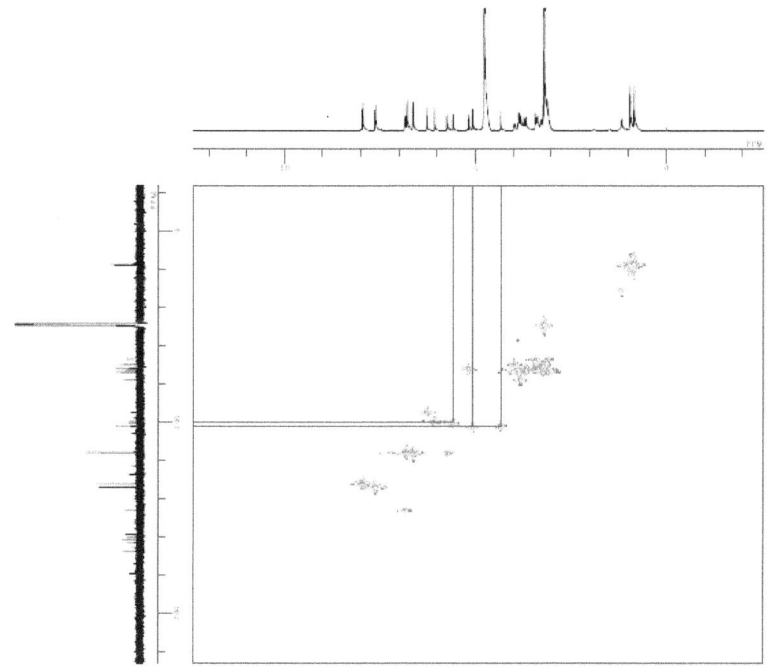

Fig. S6; HMBC correlation of compound **1** (CD₃OD, 600Mz)

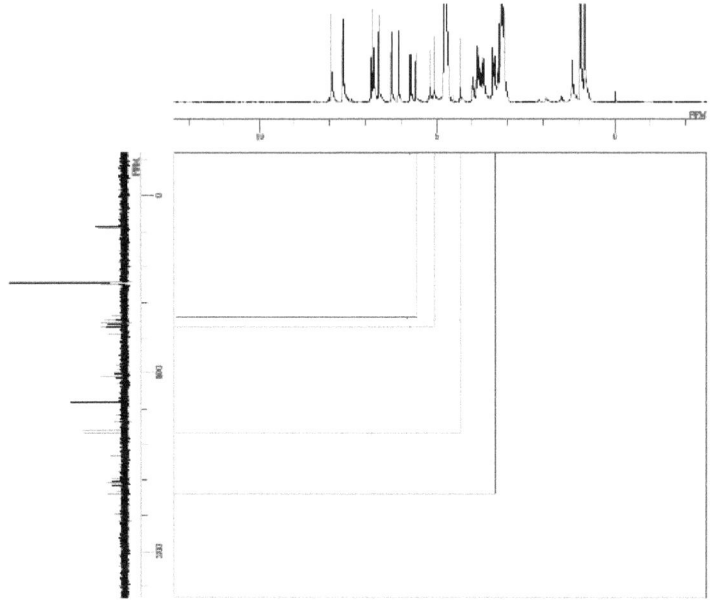

Fig. S7; HRFAB+ mass spectrum of compound **1**

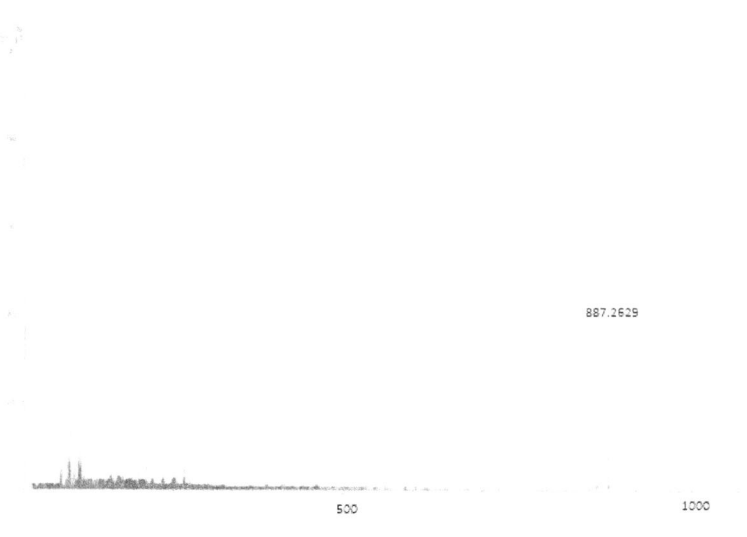

887.2629

500   1000

RT 1 60 min   Scan# 17
Elements C 42/0, H 47/0, O 21/0
Mass Tolerance   1000ppm, 5mmu if m/z < 5, 50mmu if m/z > 50
Unsaturation (U S) -0.5 - 20.0

Observed m/z	Int%	Err[ppm / mmu]	U.S.	Composition
887.2629	40.79	+2.2 / +1.9	19 5	C42 H47 O21

References

[1] R. Soare, M. Dinu, C. Bãbeanu, M. Popescu, A. Popescu. Nutritional value and antioxidant activities in fruit of some cultivars of pepper (*Capsicum annuum* L.), J. Agroaliment. Proc. Technol. 23(4) (2017) 217-222.

[2] G.E. Igbokwe, G.C. Aniakor, C.O. Anagonye, Determination of β-carotene & vitamin C content of fresh green pepper (*Capsicum annum*), fresh red pepper (*Capsicum annum*) and fresh tomatoes (*Solanum lycopersicum*) fruits, Bioscientist. 1(1) (2013) 89-93.

[3] G. Oboh, T.B.G. Rocha. Distribution and antioxidant activity of polyphenols in ripe and unripe tree pepper (*Capsicum pubescens*), J. Food Biochem. 31 (2007) 456–473.

[4] N.B.C. Prasad, R. Shrivastava, G.A. Ravishankar. Capsaicin as multifaceted drug from *Capsicum* spp, Evid Based Intern. Med. 2 (2005)147–166.

[5] Y.J. Kim, Y.A.E. Kim, T. Yokozawa. Protection against oxidative stress, inflammation, and apoptosis of high-glucoseexposed proximal tubular epithelial cells by astaxanthin, J. Agric. Food Chem. 57(19) (2009) 8793–8797.

[6] G. M. Lalanne A. J. Álvarez, M. J. Fernández, E. Azuara. Oleoresins from *Capsicum* spp.: Extraction Methods and Bioactivity, Food Bioprocess Technol. 10 (2017) 51-76.

[7] K. Sanatombi, S. Rajkumari. Effect of Processing on Quality of Pepper: A Review, Food Reviews International, 36(6) (2020) 626 – 643.

[8] Y. Kasai, K. Komatsu, H. Shigemori, M. Tsuda, Y. Mikami, J. Kobayashi, A. Cladinol. A polyketide glycoside from marine-derived fungus *Gliocladium* species, J. Nat. Prod. 68 (2005) 777-779.

[9] T. Hasegawa, F. Takano, T. Takata, M. Niiyama, T. Ohta, Bioactive monoterpene glycosides conjugated with gallic acid from the leaves of *Eucalyptus globulus*, Phytochemistry. 69 (2008)   747-753.

[10] S.R. Bornstein, H. Rutkowski, I. Vrezas. Cytokines and steroidogenesis, Mol Cell Endocrinol. 215 (2004)135 -141.

[11] A Nehmé, J. Edelman. Dexamethasone inhibits high glucose-, TNF-alpha- and IL-1beta-induced secretion of inflammatory and angiogenic mediators from retinal microvascular pericytes, Invest Ophthalmol Vis. Sci. 49(5) (2008) 2030-2038.

[12] M.C. Alley, D. A. Scudiero, A. Monks, M. L. Hursey, M.J. Czerwinski, D. Fine, B.J. Abbott, J.R. Mayo. Shoemaker R.H, Boyd M.R. Feasibility of drug screening with panels of human tumor cell lines using a microculture tetrazolium assay, Cancer Res. 48 (1988) 589-601.

[13] P.A. Segun, O. O. Ogbole, F. M. D. Ismail, L. Nahar, A. R. Evans, E. O. Ajaiyeoba, S. D. Sarker, Resveratrol derivatives from *Commiphora africana* (A. Rich) Endl. display cytotoxicity and selectivity against several human cancer cell lines, Phytother Res. 33(1) (2018)159-166.

 [14] V. Kumar, M. Mohan. Physicochemical Status and Primary Productivity of Ana Sagar Lake, Ajmer (Rajasthan), India. Univ. J. of Envir. Res. and Tech. 1(3) (2011) 286-292.

[15] P. Valentão, E. Fernandes, F. Carvalho, P. B. Andrade, R. M. Seabra, M. I. Bastos, Hydroxyl radical and hypochlorous acid scavenging activity of small Centaury *(Centaurium erythraea)* infusion, A comparative study with green tea *(Camellia* sinensis*).* phytomedicine 10, (6-7) (2003) 517-522.

[16] A. N. Panche, A. D. Diwan, S. R. Chandra. Flavonoids: an overview, J. Nutr. Sci. 5 (2016) 1-15.

[17] J. Y. Salib, H. N. Michael, E. F. Eskande, Anti-diabetic properties of flavonoid compounds isolated from Hyphaene thebaica epicarp on alloxan induced diabetic rats. Pharmacognosy Res. 5(1) (2013) 22-29.

[18] M. Halabalaki, A. Urbain, A. Paschali, S. Mitakou, F. Tillequin, A. L. Skaltsounis, Quercetin and Kaempferol3-O-[alpha-L-Rhamnopyranosyl-(1-→2)-α-L-arabinopyranoside]-7-O-α-rhamnopyranosides from *Anthyllis hermanniae*: Structure Determination and Conformational Studies, J. Nat. Prod. 74(9) (2011) 1939-1945.

[19] J.M. Calderón-Montaño, E. Burgos-Morón, C. Pérez-Guerrero, M. A. López-Lázaro. Review on the Dietary Flavonoid Kaempferol. Mini-Reviews in Medicinal Chemistry. 11(4) (2011) 298-344.

[20] A. E. De Sousa, A. C. A. Da Silva, A. J. Cavalheiro, J. H. G. Lago, M. H. Chaves, A New Flavonoid Derivative from Leaves of *Oxandra sessiliflora* R. E. Fries, J. Braz. Chem. Soc. 25(4) (2014) 704-708.

[21] W. H. Watson, Y. Zhao, R. K. Chawla, S-adenosylmethionine attenuates the lipopolysaccharide-induced expression of the gene for tumour necrosis factor alpha, Biochem. J. 342 (1999) 21–25.

[22] L. Nayely, P. Erick, G. Gutierrez, L. A. Dulce, H. Basilio, Flavonoids as Cytokine Modulators: A Possible Therapy for Inflammation-Related Diseases, Int. J. Mol. Sci. 17(6) (2016) 921-936.

[23] A. Nehmé, J. Edelman, Dexamethasone inhibits high glucose-TNF-alpha- and IL-1beta-induced secretion of inflammatory and angiogenic mediators from retinal microvascular pericytes, Invest. Ophthalmol. Vis. Sci. 49 (2008) 2030-2038.

[24] I. Crespo , M. García-Mediavilla, B. Gutiérrez, S. Sánchez-Campos, M. Tuñón, J. González-Gallego, A comparison of the effects of kaempferol and

quercetin on cytokine-induced pro-inflammatory status of cultured human endothelial cells Br. J. Nutr.100(5) (2008) 968-976.

[25] I. Okamoto , K. Iwaki, S. Koya-Miyata. T, Tanimoto, K. Kohno, M. Ikeda, M. Kurimoto.The flavonoid Kaempferol suppresses the graft-versus-host reaction by inhibiting type 1 cytokine production and CD8+ T cell engraftment. Clin. Immunol. 103(2) (2002) 132-144.

Publisher: Eliva Press SRL

Email: info@elivapress.com